EUREKA!

THE **BIOGRAPHY** OF AN **IDEA**

Glasses

BY LORI HASKINS HOURAN • ILLUSTRATED BY JOHN JOVEN

KANEPRESS

AN IMPRINT OF BOYDS MILLS & KANE

New York

For Dr. Albert, who has been helping me
see clearly since I was a kid —LHH

To D.G., for being the light to my eyes —JJ

Special thanks to Tony Chahine, OD, the Ocular Heritage Society

Kane Press
An imprint of Boyds Mills & Kane, a division of Astra Publishing House
kanepress.com
Printed in China

Library of Congress Cataloging-in-Publication Data
Names: Houran, Lori Haskins, author. | Joven, John, illustrator.
Title: Glasses / by Lori Haskins Houran ; illustrated by John Joven.
Description: First edition. | New York : Kane Press, an imprint of Boyds Mills & Kane,
[2021] | Series: Eureka! | Audience: Ages 4-7 | Summary:
"A nonfiction 'biography' of glasses, an everyday object that has become ubiquitous,
starting with the discovery of the magnifying properties of glass through
the development of the eye chart, plastic lenses, and contact lenses"
—Provided by publisher.
Identifiers: LCCN 2020052229 (print) | LCCN 2020052230 (ebook) |
ISBN 9781635924244 (hardcover) | ISBN 9781635924251 (paperback) |
ISBN 9781635924732 (ebk)
Subjects: LCSH: Eyeglasses—Juvenile literature. | Vision—Juvenile literature.
Classification: LCC RE976 .H68 2021 (print) | LCC RE976 (ebook) |
DDC 617.7/522—dc23
LC record available at https://lccn.loc.gov/2020052229
LC ebook record available at https://lccn.loc.gov/2020052230

10 9 8 7 6 5 4 3 2 1

GLASSES—THEY'RE EVERYWHERE!

Most people wear glasses to help them see. Some people wear them just for fun.

We're so used to glasses that we hardly think about them. Where did they come from? And what on earth did people do without them?

AROUND 45 CE

Two thousand years ago, a writer named Seneca had a problem. He loved to read. But the words in his scrolls were blurry.

Then Seneca noticed something. He put a scroll behind a glass of water. The words looked bigger and clearer!

He was thrilled. Now he could read all the scrolls in Rome!

Seneca's glass of water acted like a **lens**.

A lens is the piece of glass in a pair of eyeglasses—the part that goes in front of your eye.

There is also a lens *inside* your eye. Its job is to make everything you see nice and clear.

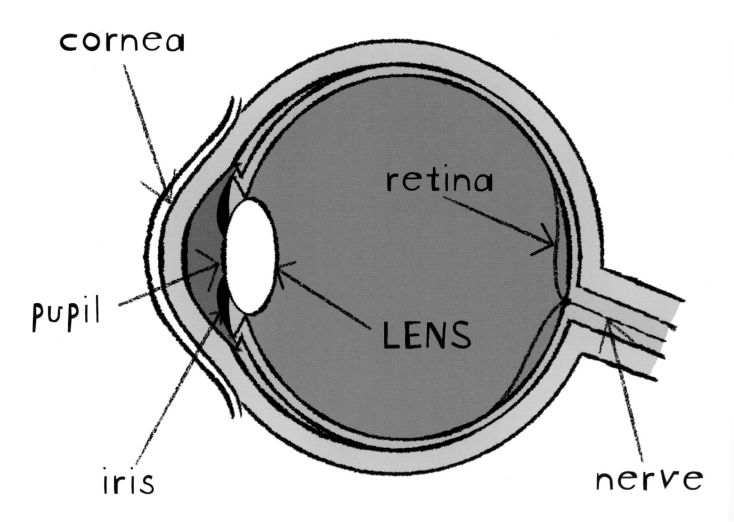

cornea

retina

pupil

LENS

iris

nerve

If some things look blurry, you might need glasses. The lenses in glasses help the lenses in your eyes work better.

AROUND 1000 CE

In Egypt, scientist Ibn al-Haytham made a discovery. A curved piece of crystal made things look larger—just like Seneca's glass did!

Word spread. People began to use crystal lenses to help them read.

Monks in England called them "reading stones." The stones had flat bottoms for sliding over pages.

AROUND 1300

The best glassmakers lived in Italy. *Why not make reading stones out of glass?* they wondered.

It worked!

One Italian craftsman—no one knows who—had another smart idea. He put two reading stones in frames. Then he linked them together. Now people could hold lenses right in front of their eyes.

It was the world's first pair of glasses.

AROUND 1440

At first, glasses weren't that popular. Most people didn't know how to read. And books were rare. Only important people like kings and scholars owned them.

The printing press changed that. It could make lots of books, fast!

More and more people bought books. More and more people learned to read. Many of them needed glasses!

AROUND 1450

Craftsmen got better at making glasses. They used wood, bone, horn, and iron for frames. They made different kinds of lenses. The first lenses helped **farsighted** people, who had trouble seeing things up close.

Later, craftsmen made lenses for **nearsighted** people,
who needed help seeing faraway things.

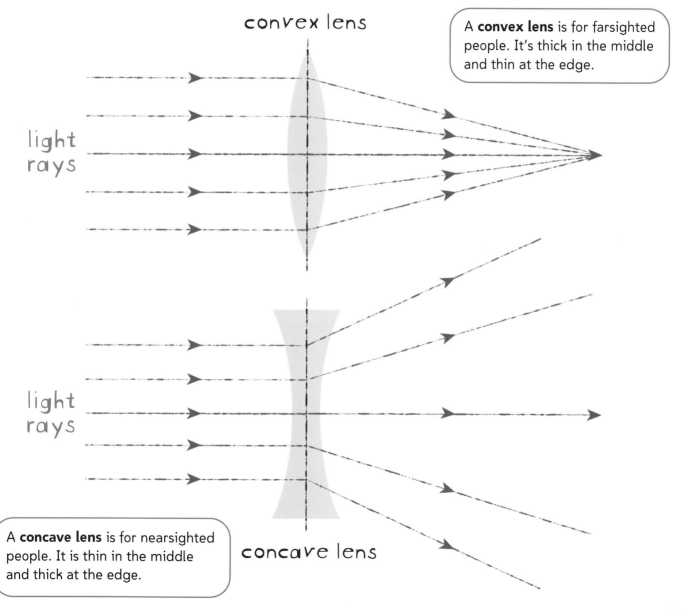

convex lens

A **convex lens** is for farsighted people. It's thick in the middle and thin at the edge.

light rays

light rays

A **concave lens** is for nearsighted people. It is thin in the middle and thick at the edge.

concave lens

• • • MORE ABOUT LENSES • • •

How do lenses work?

Let's say you look at a dog. Rays of light bounce off the dog and enter your eyes. The lenses in your eyes **focus** the rays—bring them together—to make an image of the dog.

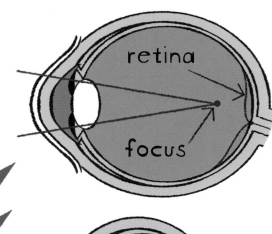

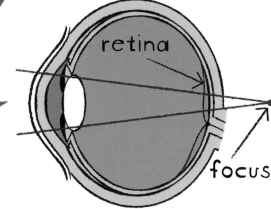

But if you're nearsighted, the rays focus in front of your retina. If you're farsighted, the rays focus behind your retina. Either way, the dog looks blurry.

Here's how glasses help. Eyeglass lenses bend the rays of light so that they land on the retina. Concave lenses spread the rays farther apart. Convex lenses bend the rays closer together. Both lenses focus the light at just the right spot.

Now the image of your dog is nice and clear!

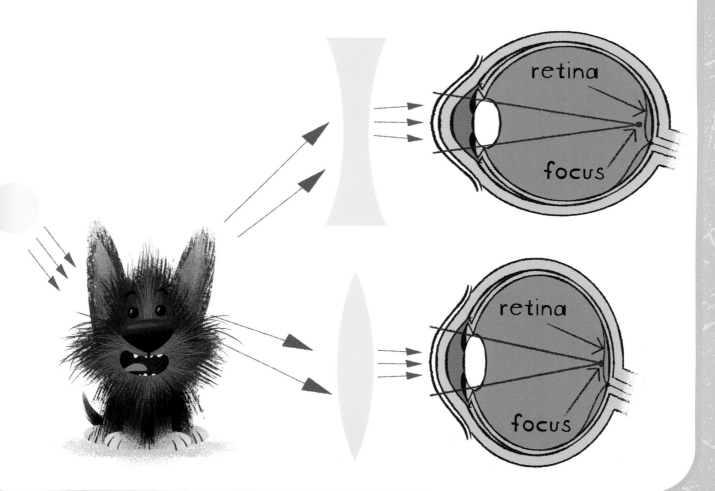

AROUND 1700

Glasses had come a long way. But there was still one big problem— how to keep them *on*!

Some glasses clamped onto people's noses. (Ouch.)

Other glasses looped over people's heads.

In Spain, people tied their glasses on with strings.

In China, they added tiny weights to hang behind their ears.

None of these things worked very well.

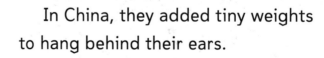

AROUND 1740

Finally, an Englishman named Edward Scarlett tried something new. He made glasses with **temples**. Temples were straight arms that rested on people's ears.

At last, there was an easy way to wear glasses!

AROUND 1784

One famous American, Benjamin Franklin, had trouble seeing both near and far. He got tired of using two pairs of glasses. So he invented **bifocals**. Bifocals have half a lens on the top for seeing far, and half a lens on the bottom for seeing near.

They're two pairs of glasses in one!

AROUND 1862

Back in Franklin's day, people didn't get glasses from an eye doctor. They stopped by the general store. They tried on different glasses until they found a pair that was good enough.

A Dutch doctor came up with a better way. Herman Snellen created an eye chart to test people's vision.

Does it look familiar? Snellen's chart is still around today!

AROUND 1942

Glasses didn't change much for a while.

Then American scientists invented clear plastic during World War II. They used it to make windows for planes. Plastic didn't break as easily as glass did.

When the war ended, lots of clear plastic was left over.

What could companies make with it?

Lenses!

Most eyeglasses today have plastic lenses. They're sturdier than glass lenses. Lighter, too.

Eyeglasses weren't invented by one person. Great thinkers from lots of countries pitched in.

Thanks to their ideas, farsighted people can read books
and paint pictures. Nearsighted people can drive cars, watch
movies, and play sports.

Glasses turn the world from blurry to clear!

··· FACTS ABOUT GLASSES ···

- **Lens** is the Latin word for "lentil," a curvy-shaped bean.

- Sunglasses were invented before eyeglasses. Inuit people made goggles from antlers and ivory to protect their eyes from the sun.

• Instead of glasses, many people wear contact lenses—thin lenses that sit right on their eyes. Did Leonardo da Vinci dream up contacts? He drew a sketch that looked like one way back in 1508!

• Roni di Lullo invented Doggles—glasses for dogs! They're mostly used to shield dogs' eyes from sun and wind, but they can correct a dog's vision, too.

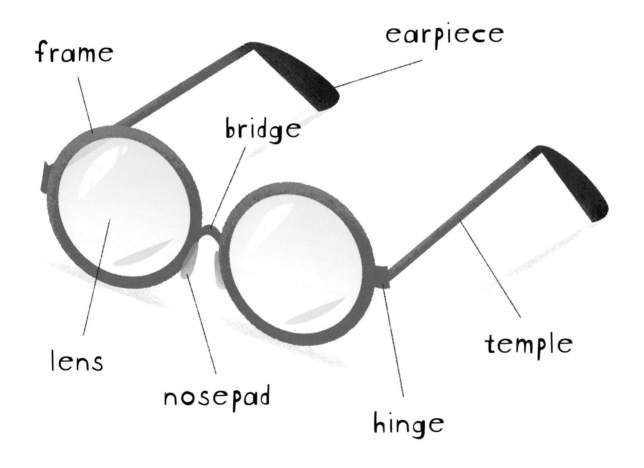

frame

earpiece

bridge

lens

nosepad

hinge

temple